EMERGING TRENDS IN SOCIAL MEDIA

Social Media Reality Of Todays Youth

Nimesh Dhakan

NOTION PRESS

NOTION PRESS

India. Singapore. Malaysia.

ISBN xxx-x-xxxxx-xx-x

Contents

Foreword

Book is about the emerging trends in social media and the social media reality of todays's youth in new media and primarily on social media.

Social media platform are the most visited and used platforms for communication and spreading information on various topics such as art , science , music , literature , sports e.t.c . As it's a thing that has been a prominent and potent tool in commumnicating and it plays a major role in marketing business activities as its now a potent tool in marketing mix and plays also a major role in spreading business in global platform.

New media today has emerged and given a boon to the social media platforms in spreading information , evolving marketing strategies and to become a potent tool in marketing mix , alternative to traditional media and also to have a better prspective in the given media format.

Nimesh Dhakan

17/12/2018

Preface

At this stage, the information era started gaining momentum and slowly reached out to every remote location in the country. The late 90's witnessed a great revolution in Information Technology. Satellite TV replaced the traditional TV broadcasting system, people could get access to more number of channels and get global news and information in real time. Today, the new media world has been completely transformed into a "high tech portal system". It has grown rapidly in terms of accessibility, context relativity, better reception, presentation styles and much more.

The days are long gone when somebody asked whether somebody is awake or asleep but in today's modern world the only thing people care about is whether one is online or offline. It is quite difficult to remain aloof and lonely with so many ways for people to connect with you. The truth is while some people are enjoying these facilities the life of introverts have become really difficult. The social media has provided people with a massive platform to share ideas, knowledge, feelings and possibly anything. Earlier you had messaging services and chatting applications but slowly photo sharing came into existence then music then videos and today what not from PDFs to PowerPoint presentations anything can be shared at an instant.This potent tools that has helped and realmed our life has given a birth to this book and also to stated a introduction to the new media platforms and primarily the social media.

Nimesh Dhakan

17/12/2018

Introduction

Trend in Real Diaspora with context to social media exist in a potential manner. They exist in almost 2 0 million people with persistence of curating free and valuable content on social medias.

With that said Social media today exist like a political hierarchy with each channel of media falling one after the another .Do they really affect the content that is being curated on all social media platforms. The content that is being curated on all social media platforms. The answer is it does effect. The Value for this content is very valuable from both perspective curators and audience. As If the chain breaks the chain and might not make the desired effect. Social media is indeed a platform to socialize in a day to day life with people.

The Diaspora will always exist of traditional media audience shift to new media. Is this the game changer for the disseminating channels to spread information and valuable free content.Today the Shift is a stalking route that enriches one mind with knowledge.It is the ability or the power of the creation that has been prominent for people to get satiated with emerging trends in this social media.

Though millennial still count it rookie platform as the content is being manipulated and revived by most of them with a feedback back on it. There is still a lot of media channel to emerge in this digital world. It is the ability of the new media transform as it technology that is dynamic and different in every walk of its creation.

The current level of media saturation has not always existed. As recently as the 1960s and 1970s, television, for example, consisted of primarily three networks, public broadcasting, and a few local independent stations. These channels aimed their programming primarily at two-parent, middle-class families. Even so, some middle-class households did not even own a television. Today, one can find a television in the poorest of homes, and multiple TVs in most middle-class homes. Not only has availability increased, but programming is increasingly diverse with shows aimed to please all ages, incomes, backgrounds, and attitudes.

1 Trend In Media Diaspora

Trend in Real Diaspora with context to social media exist in a potential manner. They exist in almost 2 0 million people with persistence of curating free and valuable content on social medias.

With that said Social media today exist like a political hierarchy with each channel of media falling one after the another .Do they really affect the content that is being curated on all social media platforms. The content that is being curated on all social media platforms. The answer is it does effect. The Value for this content is very valuable from both perspective curators and audience. As If the chain breaks the chain and might not make the desired effect. Social media is indeed a platform to socialize in a day to day life with people.

The Diaspora will always exist of traditional media audience shift to new media. Is This the game changer for the disseminating channels to spread information and valuable free content. Today the Shift is a stalking route that enriches one mind with knowledge and the information which will fill up another concept in on media stages may it be ancient old or new media. It is the ability or the power of the creation that has been prominent for people to get satiated and emerging in this new media.

Though millennial still count it rookie platform as the content is being manipulated and revived by most of them with a feedback back on it. There is still a lot of media channel to emerge in this digital world. It is the ability of the new media transform as it technology that is dynamic and different in every walk of its creation.

Media theories

It is the media theories like the Two Step Flow theory in media that has let us understand the importance of opinion leader in our day to day to life. As this is the most trending in today age that the opinion leader play important role for individual to choose their social media channel. One can always refer to any opinion leader such as reality stars can influence mass audiences to start using or delete a particular application on social media.

It is the social media in today world that satiates every individual needs in every walk of life so this will play a vital role in day to day life and also will satiate in future as this calls for new updates and also innovation or building of new app so that the cycle of any individual getting their part of satisfaction of new media can be rational enough to use and refer it to their peer group that satiates a new group.

In this emerging new media the curiosity of the audience to use a new application on play store or even give feedback on any social media platform has become alluring experience of every millennial in new age media.

Trend in real diaspora

This is the real diaspora to even inculcate every task with creativity and art in this new age media with every business getting online and also serving itself to sell their product and services to all the mass audience and also to serve them with valuable free content on all the social media channels.

This is the real diaspora to even see is the formation of all the trends in social media channels to spread from western countries to all the parts of the world is being a boggling experience. For all the new age media that has formed its roots in all parts of society and also to form the real events in this world where the space of technology could also be marked as real space that has time, space and movement of digital movement that has to be adhered to this real task that are real and also form meaning with every update in our status, blogs, stories and even audio and video devices that we upload in day to day life.

All the mass communication that is the main background in all the social media channels after e marketing and also other task in day to day life that we conquer by updates, information and other key elements of day to day life that help us to finish the task skillfully and in our daily lifestyle.

So this diasporas are just nor physical in these social media channels we get the real time diasporas that are the digital diasporas in form of audio , videos , images , information and other valuable free content.

So even one doesn't have money to reach to European country he can surely check all the platforms such as audio, video images information and also make connection with new friends in their respective country that one wants to visit and enjoy and also to get the social media updates daily. So the above example gets us to a platform of daily diasporas and also the digital importance in this golden era of new age media where one gets satiated without spending money on travelling by the digital content as well if she or she just wants the alluring experience of a sight.

This diasporas are just the day to day life experiences also that gets updated on social media platform and also the updates that one gets on RSS feeds and other daily subscription of their favorite celebrity, sports person and also the day to day to life that are the key features of this digital world that has the value of mass communication and also the boggling situation that one can remember in all day to day entertainment pinch of millions and billions of audience in day to day life.

This is the daily update the are the key feature of this social media channels and also help us to update these features in day to day life that help the needy one to get help the geek to get the information and tools the peer group to form the community that folks spreading word of mouth to reach their prospect customers.

These are not just updates some opinion leader play a vital role in influencing and also to get the mass audience to follow their opinion in all the walks of life starting from images audio and f that help us to understand that suggest that one should always be a key role that help us to involve and also to be the one that also influence to understand the key denture of day to day life that help us to civilize our self and also to get update in opinion that help us to get the maximum of content and also to update the key feature in day to day life as they play a vital role in fashion , cultural and also other diasporas in day to day life that gets us to the next level of this diasporas topic is the cultural diasporas.

The cultural Diasporas

This diasporas are the key feature that help us to learn the trends in cultural community and also the life that help us to develop and to sustain in these diasporas are as follows :-

The fashion culture

Fashion plays a vital and most important role to suggest to which generation we belong as it's not just our demographic or geographic location that suggest from the wear plain clothes to the western folks that wear these hoodies to caps to these new plain culture of Asians to wear the familiar clothes as their respectful influential leader suggest and also the one which think that fashion and fads are mostly suggested by their music choices such as hip hop , trap to folk to western to the ethnic wear that are the one task of day to day life.

The music Culture

These is the main culture that confirms of these day to day life as these are the genres which are the trap , hip hop , trance psychedelic to the western folk , trance to the modern day bass that are the daily life of the mass audiences.

The hippy Culture

This culture is not just existed in all the cultures but has just been evolving from generation to generation from every mixes that one has in all the opinions that will be depended in these types of drugs and also the key grasses that have been smoked in vapes and paper.

The technological Culture

The above are the environment that define the main criteria of a life that also explains not only day to day life but suggest the decade as well that are the culture of technology at the time of apple and Microsoft becoming the giants to the decade where the audience where literally rating photos of their peers group in order to get satiated by their entertainment that showed us the change from ancient media to old media to new media and how the content was changing.

These gets us to the individual Diasporas that are the key effects that are the most influential that also suggest that are the trends in society and culture

Top Trends in Society & culture

1. Speeding up
2. Anxiety
3. Demographic change
4. Global and local
5. Happiness
6. Authenticity
7. Memory
8. Networked
9. Us and them
10. Personalization

Top Trends in Society & culture

1. Speeding up

Everything is speeding up thanks to our obsession with technology and efficiency - although whether anything is actually moving in the right direction is a moot point. You can blame computers, email, the Internet, globalization, mobile devices, low cost travel, whatever you like. The result is 24/7 access to goods and services, multi-tasking, meals in minutes, hectic households, microwave mums, meals on the run, insecurity, one minute wins and individuals (and organizations) that want everything tomorrow. The result is stress, anxiety, a lack of sleep, a blurring of boundaries between work and home, work-life imbalance and, conversely, an interest in slowing things down.

2. Anxiety

There are approximately 40 wars in 35 countries going on as you read this. Terrorism is rife and if 'they' don't get you a global pandemic probably will. Post 9/11 the feeling was probably fearful but this has now settled down to anxiety and, if all goes well this might level off to people being slightly rattled. But the general feeling isn't going away. Trust has all but evaporated (people don't trust institutions like government or the police any longer) and the speed of change, together with technology that disempowers, has left people yearning for the past. This insecurity is to some extent generational but whether you're eighteen or eighty there is a general feeling of powerlessness.

3. Demographic change

Demographics is the mother of all trends (or, as someone more eloquently once put it, ('demographics is destiny'). The big demographic shift is ageing. In Europe 25% of the population is already aged 65+. Linked to this is the rise in single person households (46 million in Europe) caused by an increase of widows and widowers, but also caused by more people getting divorced and by people marrying later or not at all (42% of the US workforce is unmarried).

4. Global and local

Globalisation is obviously a huge trend but if you look forward far enough it looks like the future will be local. You can already see evidence for this shift in the fact that the opposite, localisation - is a major trend in everything from food to politics. And it is entirely possible that the EU could collapse back into local units or even small city-states and the consequences of this would be extraordinary. Theoretically, globalisation still has many years to run (and will run alongside an interest in all things local) but we are increasingly at the mercy of resources.

5. Happiness

Materialism is still in full swing but for many people it's starting to lose its appeal. We are working harder and working longer - and earning more money as a result - but it's becoming increasingly obvious that money can't buy you happiness. People are also starting to realize that identity is not shaped by what you own or consume but by who you are and how you live. To some extent the happiness phenomenon is really a search for meaning.

7. Memory

We increasingly live in a world that forgets. Companies have almost no sense of their own history while politicians positively revel in the fact that voters cannot remember (or choose to forget) lies, deceptions and even criminal behavior. This is a problem because power is essentially a battle between memory and forgetting. Unfortunately, memory loss is a by-product of trends like speeding-up and convergence.

8. Networked

They used to say that when the US sneezes, the rest of the world catches a cold. These days we all get to see and hear that cold in real time. Everything from countries and computers to industries and gadgets are increasingly linked together. In the future you can expect to see this trend accelerate even more thanks to everything from RFID tags to smart dust. This is both good news and bad. It's good because information (good and bad) will travel around the world instantly. This means everything becomes transparent. It's bad because in the future there will be little or no privacy and, since everything is connected, if something fails in one area the whole 'network' can be affected ('cascading failure' is the term used by some people). This explains how SARS can travel around the world at such speed and also how innovations are copied so quickly.

9. US and them

It would obviously be too simplistic to carve up the world between America (and its allies) and the rest of the world, but some people see it that way

10. Personalization

Globalization creates commodification and homogenization, which in turn creates the counter trend of personalization as people react against standard issue products. Add a dose of technology and hey presto you've got a product that users can tailor to their own tastes and needs. Expect dozens of products in different markets to offer a similar degree of personalization in the coming years as customer desire meets technological possibilities. These are just some of the trends that has been involves in our day to day life that help us to evolve and also to trend in day to day life that help us to understand the context the key feature that help us to understand the key feature that will evolve in day to day life.

"Diaspora" has become an increasingly trendy concept throughout the academic world. This is not surprising, given the incessant movement of peoples from one country, region, or continent to another for a variety of reasons: economic, political, social, and cultural. This phenomenon has called into question the relevance of the ideal-type of the "nation-state," or, more exactly, of the congruence of nation and state, and has created a situation where the societies of most countries are becoming multiethnic, multicultural, multiracial, and pluralistic. Minority populations were once referred to as refugees, immigrants, expatriates, asylum seekers, or guest workers; these categorizations seemed to be sufficient, for how else could one explain the fact that most specialists on nationalism, ethnicity, and even migration did not deal with diaspora as a distinct category or mention it at all, at least until very recently within their host countries. In any case, today many, if not most, of these categories tend to be conflated and put under the single rubric of "diasporas."

It seems clear that diaspora, once an object of suspicion, has become one of fascination. Whereas once diaspora was a historically and politically loaded concept, today it is a not only a neutral term but a catch-all one. In consequence, it is possible to put many of the above-mentioned categories under the single rubric of diaspora. This development reflects a conceptual proliferation that does not make the work of specialists easier.

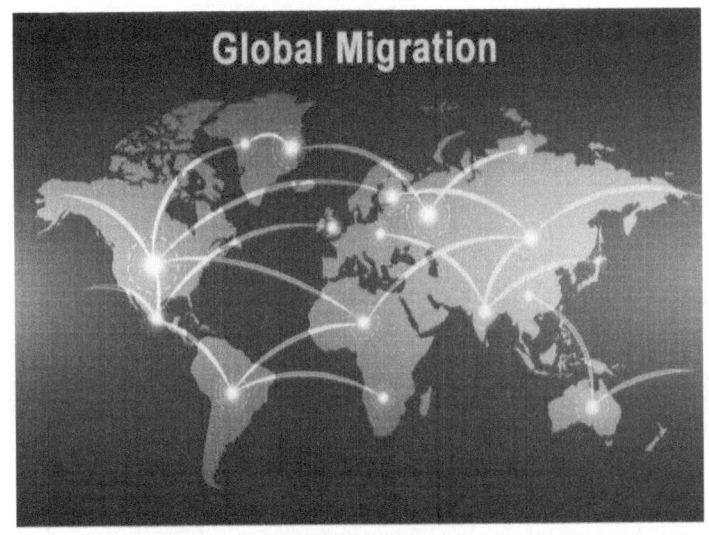

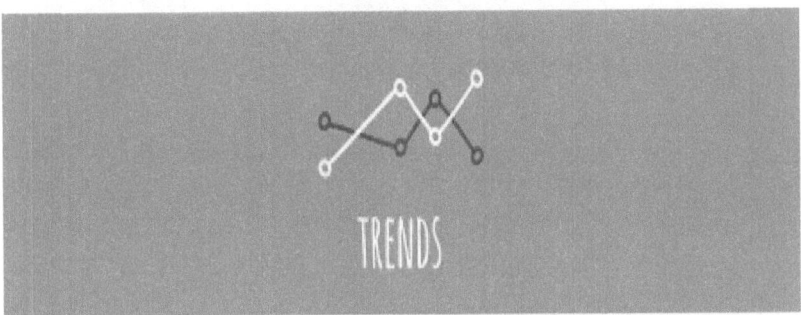

It is worth considering seven potential trends involving global migration:

1. **Proliferation of border control and immigrant identification technologies**, to track not only flows across borders, but also activities of resident immigrants. Increased use, maintenance of data bases for residents, citizens for access to services. There will likely be a related increase in opportunities for corruption, cyber intrusions, and false documentation. Technologies could give

governments capabilities they really don't want to implement, especially for large informal economies. Workarounds will abound.

2. **Sharp increase in emerging economies as immigrant destinations.** Labor migrants will take advantage of vibrant economic growth and large, urban informal economies, even if the environments portend social stresses. Governments grapple with how to accommodate immigration as both a source of economic growth and of social tension. Efforts to introduce gradations in immigrant citizenship status (as in Roman imperial efforts to give legal status to peoples from the periphery). Where will middle class interests come down?

3. **Aging societies will find ways to make labor migration work**. Aging populations and mismatches between education and labor demand will make labor migration more important to economic performance. In these aging societies, private sectors will likely sustain and increase demand for migrant labor—for both low-skill and high-skill or professional workers, even if politically and culturally sensitive. Despite episodic efforts to rein in migration, governments will generally be both unable to withstand private sector influences favoring migration and unable to systematically track and regulate individuals migrants. Are backlashes inevitable?

4. **Intensified debate over status of labor immigrants and refugees in advanced social welfare states.** We should expect increased social mobilization, legal maneuvering and NGO activities over rights and obligations of immigrants. How immigrants relate to preexisting social contracts will become an increasingly important issue.

5. **Tensions, frictions between government jurisdictions over migration.** We should expect to see divergent goals and incentives of national and provincial or local

governments, with increased efforts of urban jurisdictions to extract revenue from informal economies with extensive immigrant participation. Different jurisdictions will bear different kinds of costs for migration. We are likely to see increased attention to the obligations of residency, as opposed to citizenship, with lots of contention over which part of society can articulate such obligations. Educational standards for new migrants will likely be contested. Could inconsistencies between jurisdictions persist for years?

6. **Increased recognition by national and sub-national governments of reputational advantages of having immigrant rights** and "the right to have rights" (Arendt), at least for the highly skilled. National reputations will be a determinant of flows and, recruitment of talent and could increasingly seen as a factor in economic performance.

7. **Increased government-to-government cooperation over labor migration**. We could see some nascent global governance mechanisms, and increased incentives for governments to bind themselves in bilateral or multilateral institutions, conventions or protocols, in order to (1) gain leverage with domestic constituencies over migration issues, and (2) gain reciprocity from signatory nations. Implementing and monitoring such agreements will be difficult, contentious, and touch sensitivities regarding sovereignty.

Overall, this text was an easy and straightforward read with a strong policy lean. I believe her approach in policy is what allowed for this to be readable. The only issues I had surrounded her lack of engagement with previous understandings of transnationalism.

DIASPORA

Presently, the term diaspora emerges in regards to groups experiencing various forms of migration (e.g., forced, voluntary, labor), and whose consciousness concerns homelands, group histories, and transnational connections. The development of diaspora communities rapidly grows following and accompanying periods of war, colonialism, and globalism. have facilitated vast diasporas, which have resulted in destabilizing the notions of home, nation, community, and self. Diaspora inquiry compels us to reorganize rubrics of nation and nationalism, while refiguring the relations of citizens and nation-states. Therefore, the concepts of race, ethnicity, and diaspora refer to both identity and social relations. As a nation, we have tried to move beyond race, to a post-racial age, and yet we continue to differentiate, distance, mistreat, or turn a blind eye to others.

Presently, the term diaspora emerges in regards to groups experiencing various forms of migration (e.g., forced, voluntary, labor), and whose consciousness concerns homelands, group histories, and transnational connections. The development of diaspora communities rapidly grows following and accompanying periods of war, colonialism, and globalism. These facilitated vast Diasporas, which have resulted in destabilizing the notions of home, nation, community, and self.

Diaspora inquiry compels us to reorganize rubrics of nation and nationalism, while refiguring the relations of citizens and nation-states. Therefore, the concepts of race, ethnicity, and diaspora refer to both identity and social relations. As a nation, we have tried to move beyond race, to a post-racial age, and yet we continue to differentiate, distance, mistreat, or turn a blind eye to others.

These terminologies to the real meaning of these diasporas what really matters all around in these day to day life with digital diasporas that really effect in our day to day life in these walks of life right from the peer group to other cultural groups to ours friends and family that will have the most diasporas the most parts of the world.

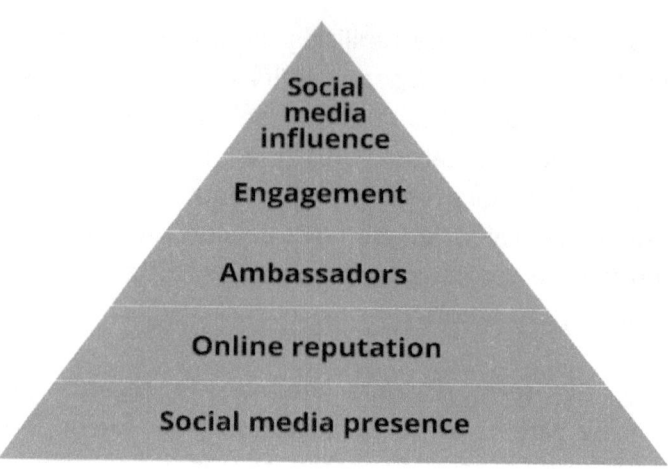

Let's just now study a popular case study of business Diasporas that has changed in our day to day life from development in these new media

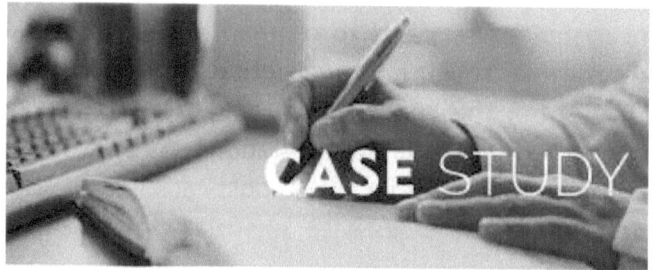

Case Study.

18-Year-old Sai Reddy makes over 12300$ a week from new mobile app.

Sai Reddy is now officially one of the richest 19-year olds in the Pune and the entire India, for that matter. The Pune native was playing on his smartphone last week when he discovered a popular new app called Olymp Trade. He had recently downloaded it but hadn't played it yet. So, Sai opened the application and started playing with the demo account you get when registering for free. In just 15 minutes he doubled his money, making more than his parents earn working hard the entire month.

One week later, Olymp Trade binary options deposited 1200$ of Sai's earnings straight into his mother's bank account. Sai is the only son of Vihaan and Pari Reddy, a typical middle-class couple, both of whom are originally from Chennai. Sai's mom taught him how to use a computer at the young age of 6 and ever since then, he had been hooked. When he's home, he can be found usually playing video games online or on Youtube.

Their family did not appear to be the "stock trading" type though. That's why his father, Vihaan was shocked when Sai showed him his balance on his phone! Mr. Reddy tells us, "I was watching the game with my wife when all the sudden, I heard Sai yell out, 'Dad, Mom, I just made 120$ for us!' I thought he was talking about some game on his computer again. I went into the bedroom to see what the fuss was all about and saw his trader account with graphs and numbers and most importantly with 130$ on it.

Diasporas have traditionally been conceived as imagined communities that maintain relatively detached ties with their countries of origin. This is about to change as a result of the global proliferation of Information and Communication Technologies. Digital tools now enable diasporas to remain in continuous contact with their countries of origin. Likewise, governments use digital tools to strengthen political ties with global diasporas. Thus, through the digital world these imagined communities have transformed into virtual communities.

The transition from imagined communities to virtual ones is of great relevance to development studies. States, on their part, use digital tools in order cement their relations with virtual diasporas, incorporate them in national advocacy campaigns and seek their assistance in creating political and financial ties with foreign countries

2. Media A Boon In Disseminating

Day to day life Media : Boon or Bain ? Lets argue

From Television to Newspapers, from Radio to Internet, media plays an important role in every sector of the entertainment industry. Though media has been criticized for their bold and straight-forward approach in the society, we cannot neglect the advantages that it brings with it. I still remember few incidents, which were initially neglected, but gained popularity with the ongoing news channels, and newspapers. Exam paper leaks, robbery and various other crimes, are informed by the media, with the help of television, newspapers, Internet and radio.

Some of the **Advantages** of media are:

1. **Quiz Programs**: The various quiz programs like KBC, are very informative and educational for students like me. We get to know various facts and also increases our general knowledge.

2. **Internet**: Known as the world wide web, Internet has raised the bar of technology over the years. It has reduced the distance between people over the years, and also increased our knowledge. Every type of news are available here.

3. **Consumer Products**: Every type of consumer products are showcased with the help of advertisements in our televisions and newspapers, which helps us to decide, which product matches our requirements.

But the above advantages, does not neglect the **disadvantages** of the media, which are as follows:

1. **Misuse of Internet**: Though Internet informs us about the latest happenings of the world; it does not always provide the 'good'

information of the society. Sometimes, spending and wasting time in the internet, can be harmful for children and students.

2. **Addiction**: Some television programs, If addicted can pose a serious problem for children. Spending too much time watching movies and serials, can affect their studies and growth.

3. **Health problems due to Addiction**: Prolonged view of Internet and TV can also cause eye sight problems, and obesity among children of various age groups.

Thus, before concluding the argument about if its boon or bain lets get into some arguments that can help us some conclusion. I would like to say that Media can be both boon and bane, depending on our usage. If we use it intelligently, it will be useful in gaining knowledge and for best impact. Also some stunts in movies, and television are copied by people, which leads to major accidents in our society. Thus, we should be sensible while choosing, and choose whatever is best for us.

These are just some points that helps us to conclude before suggesting a answer in day to day life will be helpful that will be sufficient enough to suffice the needs of our day today life that will help us to further conclude with some constructive argument.

Let's just see the advantages and dis advantages what all the folks are driving their self into and how it is disseminating those information to conclude further.

Social Media: A Boon or Bane

The days are long gone when somebody asked whether somebody is awake or asleep but in today's modern world the only thing people care about is whether one is online or offline

It is quite difficult to remain aloof and lonely with so many ways for people to connect with you. The truth is while some people are enjoying these facilities the life of introverts have become really difficult. The social media has provided people with a massive platform to share ideas, knowledge, feelings and possibly anything. Earlier you had messaging services and chatting applications but slowly photo sharing came into existence then music then videos and today what not from PDFs to PowerPoint presentations anything can be shared at an instant.

But there are always two sides of a coin and while there are the good and pleasant things there is also the negative side where the social media can be misused and have atrocious ill effects on people. Let us begin on the positive side and then trot on the negative side because beginnings are expected to be attractive.The main idea behind the social media platform was to connect people from around the whole world at just the price of a computer, laptop, tablet or mobile and an internet connection. This target of the basic idea of connecting people has been reasonably met not fully as people who cannot afford both of these still have to use the conventional ways of letters and phone calls to remain connected.But common people are greatly engrossed in this virtual world where you can have a completely different personality from your real life personality. There has been a popular mindset nowadays that the family which is connected by Facebook stays together.

Though families who are not there also do stay together but it is a trend nowadays to ask somebody if you use this particular app or not before even exchanging numbers with somebody. Nobody wants to spend a whole one rupee to send a 160 character long text message anymore.

People do not need to have those long telephonic conversations anymore after a movie or a date or provide those long stories about their vacation trips. They first let you know where they are. Now while the world is one happy place for some there are some other people who are using social media as a tool and a potent weapon to satisfy their vested interests. It is almost a daily affair where violence and cultural problems arise on a controversial tweet or even a small quote. Terrorist organizations and extremist groups use the social media to propagate their beliefs and instill their mindset and thoughts among people. The problem is that because of the audio visual power and the spread of social media these unsocial elements are being successful in doing this and spread their poison across countries at large.

In conclusion, social media has its pros and cons and the power of social media can be dangerous at times so we ought to be aware and careful to not let the negative side of the coin shine brighter than the positive side.

"Content is the present and future of the media"

Internet: Boon or Bane?

Internet has a great influence in our lives. We may or may not realize, but it has changed a lot in our lives. We are surrounded by mobile phones, computers and many other electronic devices all the time. Actually sometimes you might actually feel being a slave of these devices. These changes are both advantageous as well as not so much advantageous for us.

Connecting across the Globe

Social networks like Facebook and Twitter allow us to connect with people who, in a pre-Internet age, we might not have known about. On an average, people at facebook have 350 friends. Also, one tweet from twitter shakes the whole country, change opinions of millions of people in a second.

E - Marketing is the new field in the industry

E- Marketing is the new field and also the new one to even consider and also to suffice the business needs that can be thought in day to day life and also to sufficiently add up the things on all the dates that can be enough

Dissemination

Online Content

These content make up the new media without which there will be nothing that will be available to our day to day life that will suffice our needs in our day to day life.

Technology has also promoted the phenomenon of "selfie." There are sites sharing information on "how to have a perfect selfie" and "different poses for selfie." New terms have been introduced such as koolfie, restaurantfie, musclefie, dentisfie, and many more. Introduction of "selfie sticks" and "selfie shoe" have enhanced obsession among people for selfies.

Selfies are never reported as an official cause of death. It is believed that selfie deaths are underreported and the true problem needs to be addressed. For example, certain road accidents while posing for selfies are reported as death due to Road Traffic Accident.

Thus, the true magnitude of problem is underestimated. It is therefore important to assess the true burden, causes, and reasons for selfie deaths so that appropriate interventions can be made. Previous studies have taken data from Wikipedia and Twitter images, which may underestimate the true number of selfie deaths. Our study incorporates the records from news reports, thus we were able to get a wider range of information related to selfie deaths.

These trends are just mere seeds that get sprouted by every social media giant and then it gets budding and help in growing and also further embraces its effect and also how audience interpret their culture on in order to form the meaning and also to convey a meaning and create a culture.

To talk about advantages, there are plenty of them. One such advantage of e-commerce is very lower transaction costs. By serious automatic processing, the web can appreciably lower both taking- order costs and service to the customer costs after the sale. Another great advantage is that online shops run 24/7 and never ever close. A site called Amazon.com offers a unique feature. It includes the option of checking out the related books that people purchased after buying the book which the user wanted. Here you can see interconnected books that people are buying which a normal bookstore cannot offer, A well incorporated website can lure customers with more information than previously obtainable. Other great advantages are the facility to put up order over several days, facility to see actual prices of products by organizing products, facility to effortlessly build custom orders which can be very complicated, facility to easily discern prices between more than one trader and the facility to explore.

Technical

some technological solutions could be expensive though it's believed that e-commerce is cheap. Some protocols required are not quite standardized around the world. There is not that much reliability on some processes. Insufficient telecommunications bandwidth and limited access to dial up; cable, ISDN, wirelesses are also some of the limitations.

Non-Technical-Limitations

There is a fear to the customers that whether their personal information given is wrongly used. Also, outsourcing technical support in foreign languages is difficult. Most of the people are not used to paperless, faceless, non physical transactions. So trust factor is less

Dissemination through the media

One aspect of dissemination is media coverage of your research. This can feel flattering and exciting, and can be very valuable for dissemination. But there are important ethics considerations in disseminating research through the media.

Research findings may be picked up by press in a variety of ways. Correspondents from national and local news media often attend conferences in order to learn about new research findings, or your funder or institutional press office may release information about your research findings (although they will usually tell you before they do this). But journalists also use to the internet to identify researchers that could comment or provide an expert perspective on a story they are developing, and enquiries sometimes come completely out of the blue.

Your organization may offer training and support in working with the media. Press officers (in your institution or in the funder's press office) can be very helpful with dissemination, but this can also necessitate some care. When someone else is summarizing your research, there is always a potential risk that they could present simplistic, sensational, or inaccurate reports. Work closely with them, and check press releases before they go out. Also often offer advice and support. If you think it is possible that your research may attract media interest, it is very worthwhile to spend some time doing this kind of training. They are aimed at psychologists, but are more widely relevant.

At the very least, it is worth spending some time talking to colleagues who have experience of working with the press. There are certainly some horror stories out there – of research findings being misrepresented, and of reputational damage to individuals and institutions – but media coverage can be helpful, if you know how to get the most out of it.

It can be very helpful to work closely with a research-friendly journalist (i.e. one who understands research), who will allow you the opportunity to look at what how your work has been presented before going to press. If you are keen to get your research disseminated more widely, you will need to spend some time and effort in developing these relationships.

Before you go on and respond to the media, we suggest you work through the following key points:

Do you need to tell anyone that you are speaking to the press? Your organization or your funder may require that their press offices check and approve any press releases reporting on findings from your study.

Trends that are emerging

Emerging Trends In Social Media

At this stage, the information era started gaining momentum and slowly reached out to every remote location in the country. The late 90's witnessed a great revolution in Information Technology. Satellite TV replaced the traditional TV broadcasting system, people could get access to more number of channels and get global news and information in real time. Today, the media world has been completely transformed into a "high tech portal system". It has grown rapidly in terms of accessibility, context relativity, better reception, presentation styles and much more.

Technology has indeed had a major impact on the media trends being set today. High Definition Video has become the norm & integral component, starting from D2H TV to on-line TV channels. The second notable development is the increased accessibility. This happened through Internet TV. The on-line channels can now be viewed from a simple PC to advanced mobile Devices from any remote location. In addition, the existing and emerging Media channels have substantially increased their quality levels to meet the global standards. The average viewers are being exposed to innumerable choices which never existed earlier. The percentage of on-line Media viewer-ship is steadily increasing since 2008. Today, one in every 5 mobile users in India is using the Mobile technology to get news updates and this ration is fast increasing.

The greatest motivating factor has been the spectacular rise of regional language channels. The true era of Information revolution is emerging now. The media is finally coming out of its imposed language limitations. The new approach is to make the media more viewers friendly. With launch of many regional channels for News, Entertainment and Life style, viewers from the remote corners of the country where people still prefer their regional languages over others, have been increased drastically.

The social media has grown phenomenally in the last decade from being just a "friendship development platform" Almost every media channel today has a dedicated web portal that connects many global social media websites. More than 35% of the news are generated through social media. It is also a center stage where the average viewers write/express their views and opinions about many current events freely. This helps the Channels & Media Houses in making their content more context sensitive and understandable.These news are also from the positive News Network that help us to conclude and also to use it in a day to day life that help us in a number of things that sums up a number of things that really helps to conclude in a day to day life that is really that sums up the rest of the news that are generated via Agenda Setting theory.

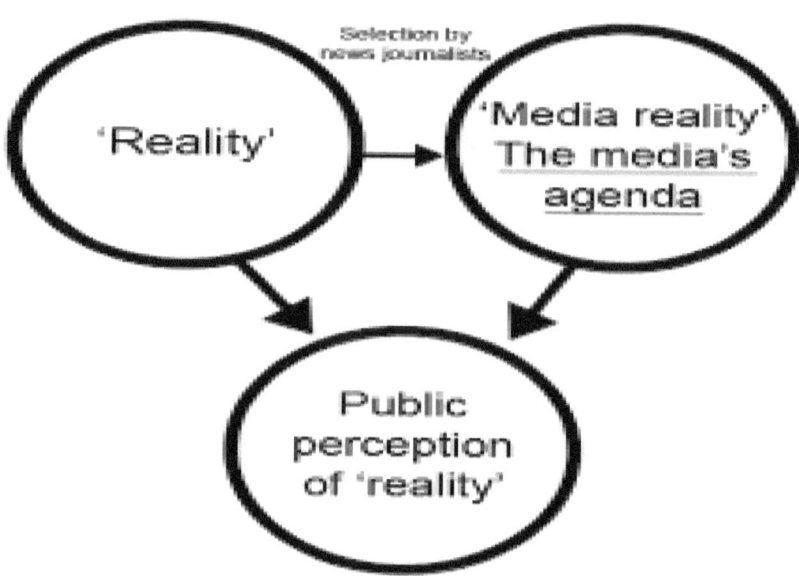

3.Trends Emerging in Media

Agenda Setting Theory

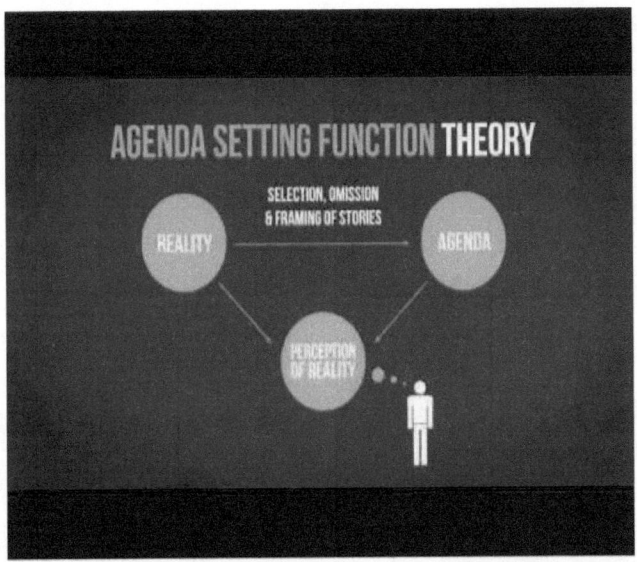

Further It suggest that the reality that gets selected by journalist is then converted into media reality which then gets converted into Public Perception of reality.

Agenda setting describes a very powerful influence of the media – the ability to tell us what issues are important. As far back as 1922, the newspaper columnist Walter Lippmann was concerned that the media had the power to present images to the public. McCombs and Shaw investigated presidential campaigns in 1968, 1972 and 1976. In the research done in 1968 they focused on two elements: awareness and information. Investigating the agenda-setting function of the mass media, they attempted to assess the relationship between what voters in one community said were important issues and the actual content of the media messages used during the campaign. McCombs and Shaw concluded that the mass media exerted a significant influence on what voters considered to be the major issues of the campaign.

Agenda setting theory (Maxwell McCombs and Donald L. Shaw)

Media influence affects the order of presentation in news reports about news events, issues in the public mind

These are the levels of agenda setting theory:
First Level:
Mostly studied by researchers, media uses objects or issues to influence the people what people should think about.

Second level:
Media focuses on the characters of issues how people should think about.

The main concept associated with the agenda setting theory is gate keeping. Gate keeping controls over the selection of content discussed in the media; Public cares mostly about the product of a media gate keeping. It is especially editors media itself is a gatekeeper. News media decides 'what' events to admit through media 'gates' on ground of 'newsworthiness'.

For e.g.: News Comes from various sources, editors choose what should appear and what should not that's why they are called as gatekeepers.

To say in simple words, Media is giving utmost importance to a news so that it gives people the impression that is the most important information. This is done every day the particular news is carried as a heading or covered every day for months.

Headlines, Special news features, discussions, expert opinions are used.

Media primes news by repeating the news and giving it more importance E.g. Nuclear deal.

Framing
Framing is a process of selective control

Two Meanings

1. Way in which news content is typically shaped and contextualized within same frame of reference.
2. Audience adopts the frames of reference and to see the world in a similar way. It is how people attach importance to a news and perceive it context within which an issue is viewed.

Framing talks about how people attach importance to certain news for e.g. in case of attack, defeat, win and loss, how the media frames the news such that people perceive it in a different way.

We can take India and Pakistan war; same happening is framed in different ways in both the countries. So depending on which media you view your perception will differ.

EMERGING TRENDS IN NEW MEDIA – THE WAY FORWARD FOR BUSINESS AGILITY

New Media is the fastest-growing segment of the Media and Entertainment industry.

It encompasses all next-generation forms of communication, including digital publishing platforms and online gaming.

In order to power growth, New Media companies must ensure personalized services by investing in data analytics. The challenge for these outlets is to balance ad load and viewer engagement thereby maintaining best-in-class viewer experience.

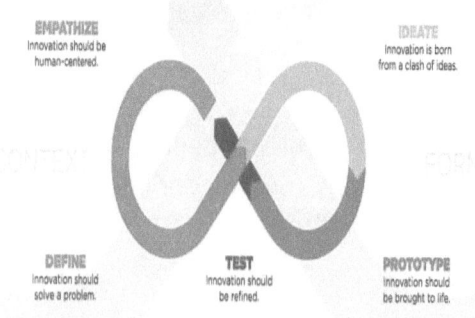

Design Thinking

This are the basic in implementing ideas on new media such as social media platforms that are trending in today's age.

Multipurpose Platforms

Multipurpose platforms are very much important to convey many things that are that are the key to disseminate information that help us a no of work are important in many information and also to convey a no of number of information.

Trends in New Media

Following are the technique which are trending in social media

In 2017, about 80 percent of consumers said that when it comes to branded content, they'd rather watch a live video than read a blog, and 82 percent prefer live video to written social media updates. And according to Facebook, live videos receive six times more engagement than non-live ones.

According to Social Media Examiner, 61% of marketers plan to increase their uses of live video going forward. As for what to broadcast, many brands are finding success by live streaming events, which helps brings an event to followers that they might otherwise not be able to attend.

2. Instagram Stories is Crushing Snapchat

Instagram launched Stories to compete with Snapchat. It features a series of bubbles at the top of the app that show users' shared photos and video clips for up to 24 hours, and has filters too. By summer 2017, Instagram Stories had 250 million users compared to Snapchat's 166 million.Today, about half of the businesses on Instagram produce a story each month, and 1 in 5 organic Instagram Stories from brands see at least one direct message from a consumer.Instagram Stories makes it easy to reach users where they already spend time, and brands will be incorporating this feature into their social strategies.

Emerging Trends in New Media

Marketing

The days of 'Interruptive marketing' are numbered. For example, 71 per cent of US consumers have recently purchased a 'time-shift' box in order to skip the adverts on television. Direct marketing is rapidly becoming a media dinosaur, soon to be extinct. In future search engines could flag-up trusted sites, as a way to help consumers avoid those subject to pop-ups or those with a less favorable reputation. With television and print advertising revenues dropping dramatically over recent years, and only radio revenues rising (people can do other things while they listen to a radio) it would seem that ambient adverts, such as posters on bus shelters, will be on the increase, since the viewer has no control over them. In fact some retailers, such as Thomas Cook and Audi, have themselves become broadcasters, and in the case of the former, have their own channel, deemed cheaper to run than paying for advert space amongst more traditional broadcasts.

The industry 'think-tank' Demos recently did a survey which questioned 3G phone users whether they liked their phones. All replied enthusiastically in the affirmative.It then informed them that their service providers knew exactly where they are, when their phones were on, at all times, and asked them again how they felt about their phones and this covert exchange. The response was very different and previously happy customers were lost.

Television

Within the last 18 months Pod-casting has taken off in a massive way, customers listening to their favorite time-shifted radio, and the BBC has taken note of this and are currently working on video on demand where viewers will be able to watch what they want to when they want to. The only limitation to this integrated media player being that requests will probably only be available at least seven days after the date of initial transmission.

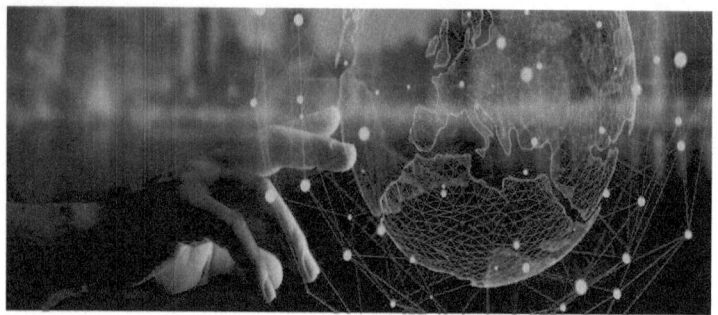

The Web

SeeMeTV demonstrates the high demand out there for user-generated content and for sites where the customer is involved in the creation of the site, providing their own content, like blogging, but instead of words, pictures and videos are sentMusic and entertainment sites are at the forefront of consumer research and targeting their desired audiences. Since many young people receive all their functionality from these sites it is perhaps time for other companies to take the time to learn from them.

trend

A pattern of gradual change in a condition, output, or process, or an average or general tendency of a series of data ...

General trends

In future people will pay for their services not solely by cash but by information exchange; providers will monitor spending habits and pass the information on to other interested parties for further remuneration.Technology is also changing the way in which people are educated and more and more revolves around specific information and the ease of how to get it.

Voice Search Will Change The SEO Industry

Imagine a world where you have multiple Alexa devices in your home. You're planning your movie night in and you say, "Alexa, order me a pepperoni pizza." Alexa directs your request to Google's search engine and orders a pizza for you.The future of search for marketers will drastically change because we will need to figure out how to build applications to the devices. The importance of organic content-writing on websites will become less valuable to consumers because the in-home device will be retrieving the information and serving it to the user.

Development Will Start With A Mobile-First Approach

Today we see more mobile traffic to sites than desktop, and we use this data to drive our design and development strategies. As consumer behaviors shift, so should the thought process of marketers. Pay attention to people and understand behavior. If you can understand and predict the way people are moving, you can develop strategies and make decisions that are driven by data and not by intuition.

Video Content Must Capture People's Attention

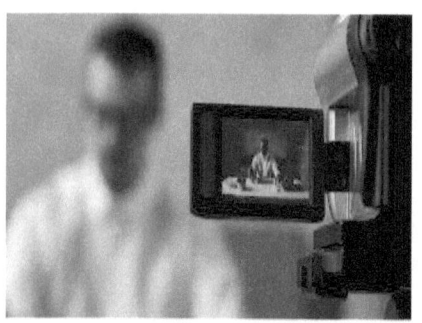

Social media platforms have altered their algorithms to favor video content. Now, social users are seeing fewer photos first and immersing themselves in the black hole that is the endless scroll of a video playlist. One tactic that beats news feeds and provides long-term value for your business is Facebook Live events. Facebook live events not only notify all of your followers to attend the live event, but they save the Video for future views.

In mass communications programs, you study different communication theories that try to explain how people listen, understand, and think about what others say to them. Depending on the program, you can focus your studies on journalism, advertising, public relations, or even radio and TV broadcasting. Some mass communications programs focus more on the effect of mass media on society, and its role in today's world.

Elaborations of Mass Media and Emerging Trends

Some of the features of working in Mass Media that you'll need to know.

Most professionals who work in this line of work, majority of the time start out as print reporters. Newspapers are also tangible. You can actually hold the palpable story in your hands. You can even cut it out with scissors and make photocopies, or hit cut-and-paste and forward a story to your boss and co-workers.It's much more difficult to capture an audio clip from a radio story, or a video clip of a TV piece. Stations -- especially radio stations -- broadcast all day, 24./7. And it is true, that newspapers typically break stories that TV and radio stations then cover. It's not a covert that radio and TV producers read the morning paper when they're looking for things to fill the daily broadcasts and kill sometime.

Today, people can get their news from thousands of different outlets. There are hundreds of outlets on paid TV and internet radio. You can access just about any newspaper you want online by the click of your fingertips. These days when the entire nation goes home from work, reading the newspaper and turning on the television. If one wants to reach more than a slice of the demographic that you sought for, they have to get into not only newspapers, but radio, television and the internet, etc.

Today's audience being so fragmented means any plan to get press coverage has to cover all of those bases. You can't send the same press releases to every media outlet and call it good. A release that's the right size for a newspaper is far too long to read on the radio.

4.Future of Media In Forming Popular Culture

The Role and Influence of Mass Media

Mass media is communication—whether written, broadcast, or spoken—that reaches a large audience. This includes television, radio, advertising, movies, the Internet, newspapers, magazines, and so forth.Mass media is a significant force in modern culture, particularly in America. Sociologists refer to this as a **mediated culture** where media reflects and creates the culture. Communities and individuals are bombarded constantly with messages from a multitude of sources including TV, billboards, and magazines, to name a few. These messages promote not only products, but moods, attitudes, and a sense of what is and is not important. Mass media makes possible the concept of celebrity:

without the ability of movies, magazines, and news media to reach across thousands of miles, people could not become famous. In fact, only political and business leaders, as well as the few notorious outlaws, were famous in the past. Only in recent times have actors, singers, and other social elites become celebrities or "stars."

The current level of media saturation has not always existed. As recently as the 1960s and 1970s, television, for example, consisted of primarily three networks, public broadcasting, and a few local independent stations. These channels aimed their programming primarily at two-parent, middle-class families. Even so, some middle-class households did not even own a television. Today, one can find a television in the poorest of homes, and multiple TVs in most middle-class homes. Not only has availability increased, but programming is increasingly diverse with shows aimed to please all ages, incomes, backgrounds, and attitudes.

What role does mass media play? Legislatures, media executives, local school officials, and sociologists have all debated this controversial question. While opinions vary as to the extent and type of influence the mass media wields, all sides agree that mass media is a permanent part of modern culture.

Class-dominant theory

The **class-dominant theory** argues that the media reflects and projects the view of a minority elite, which controls it. Those people who own and control the corporations that produce media comprise this elite. Their concern is that when ownership is restricted, a few people then have the ability to manipulate what people can see or hear. For example, owners can easily avoid or silence stories that expose unethical corporate behavior or hold corporations responsible for their actions.

The issue of sponsorship adds to this problem. Advertising dollars fund most media. Networks aim programming at the largest possible audience because the broader the appeal, the greater the potential purchasing audience and the easier selling air time to advertisers becomes. Thus, news organizations may shy away from negative stories about corporations (especially parent corporations) that finance large advertising campaigns in their newspaper or on their stations. Television networks receiving millions of dollars in advertising from companies like Nike and other textile manufacturers were slow to run stories on their news shows about possible human-rights violations by these companies in foreign countries. Media watchers identify the same problem at the local level where city newspapers will not give new cars poor reviews or run stories on selling a home without an agent because the majority of their funding comes from auto and real estate advertising. This influence also extends to programming. In the 1990s a network cancelled a short-run drama with clear religious sentiments, Christy, because, although highly popular and beloved in rural America, the program did not rate well among young city dwellers that advertisers were targeting in ads.

Popular vs. Mass Culture

☐ TV serial shows as an item of big-audience culture

☐ Big-audience culture — a neutral substitute for populare culture or mass culture

☐ MC ⟶ exclusive (critical) paradigm

☐ PC ⟶ inclusive (populist) paradigm

Pop Culture vs. High Culture

Pop culture is the culture of the people and it is accessible to the masses. High culture, on the other hand, isn't meant for mass consumption nor is it readily available to everyone. It belongs to the social elite. The fine arts, theater, intellectual pursuits — these are associated with the upper socioeconomic strata and require more a high brow approach, training or reflection to be appreciated. Elements from this realm rarely cross over into pop culture. As such, high culture is considered sophisticated while popular culture is often looked down upon as being superficial.

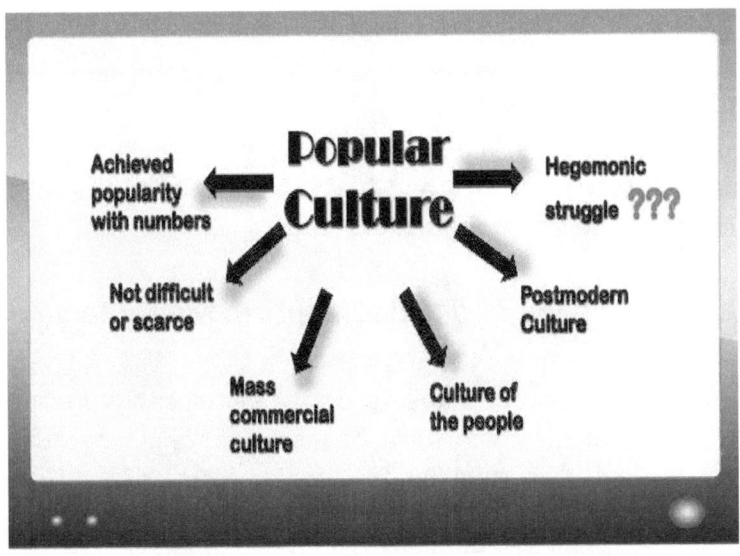

Mass Media and its Influence on Culture:

What is the first thing most Americans do when they wake up in the morning, when they get out of their car, or when they are done with work? They check their cell phones. Currently communication and media have a strong hold over our everyday lives, we search for answers in the media, looking to form opinions and learn about what in the world is affecting us personally. Through the media we decide what places we are going to go, our political parties, what is in style, and many other things that seem to be crucial factors. Mass media has developed drastically over the last century, and each change has influenced Culture. This paper will discuss both the changes and effects, and how we use the media in our everyday lives in the following three sections:

1. What were the major developments in Mass Media over the last century?

2. How have these developments affected Culture?
3. What is Media Convergence and Media Literacy and how do they affect everyday life?

Understanding media and the role it plays in society is key to forming educated opinions and having control over the media's influence and the information it presents.

What were the major developments in Mass Media over the last century?

Media has been changing drastically since its inception. While print was the main form of media for the longest time, when radio and television became major forms of media they brought something with them that words could not do. As technology expanded and matured, the electronic age began. Since that time mass media has multiplied exponentially, information can be accessed faster and faster, more and more information could be circulated and accessed. As well as information circulating, marketing and ads were now able to reach more people, and target audiences based on individuals' internet use. The internet influences the growth and spread of media, and eventually gave the power of media transfer to the people, now publishers and stations are no longer in control of the media and its information. New sources of mass media, like social networking sites and blogs are considered emerging media, because they are not from trained journalists or publishers, but the people themselves, and this development has changed media drastically in the last 10 years.

How have these developments influenced Culture?

Obviously as the ways of accessing media have evolved, so has the way it affects the individuals who have access to it. When radio and television news first became popular, most used that and the newspaper as their primary sources of information, and did not question where that information came from, or if it was valid or not. As people have begun to receive more and more information from different sources, and see that there are more points of view and opinions on the same information, this is shown in both traditional and emerging media. This new source of information is a huge effect on not only Culture, but the entire world. As people become more aware about what is truly going on, and taking about it through emerging media sources and exposing themselves to many different sources of traditional media, the individual is able to find the truth and form his or her own educated opinion, and Culture holds this as a very important factor in the people's everyday lives. We feel as if we are not droids controlled by the government, but people in control of our country, and emerging media has made this so.

What is Media Convergence and Media Literacy and how do they affect everyday life?

Throughout human history new technologies of communication have had a significant impact on culture. Inevitably in the early stages of their introduction the impact and the effect of such innovations were poorly understood. Plato used the voice of Socrates to raise the alarm about the perils posed by the invention of writing and of reading. In his dialogue Phaedrus, Plato denounced writing as inhuman and warned that writing weakened the mind and that it threatened to destroy people's memory. Also the invention of the printing press was at its time perceived as a threat to European culture, social order and morality. "Ever since they began to practice this perverse excess of printing books, the church has been greatly damaged," lamented Francisco Penna, a Dominican defender of the Spanish Inquisition. Similar concerns have also been raised in the aftermath of the ascendancy of the electronic media—television in particular has been often represented as a corrosive influence on public life. That is why for so many of its critics its impact on offline culture appears in such a negative light. There is little doubt that the digital technology and social media has already a significant impact on culture. Towards the end of the 19th century artists sough to capture their subjects through portraits of individuals who were absorbed in the act of reading a book. Today, it is the pictures of people standing in the middle of a crowd, captivated by what they are reading on their smartphone that best symbolizes the 21st century subject.

Technology and Culture

The Internet and social media are very powerful tools that can influence and shape human behavior. The social media has played a significant role in recent outbreaks of social protest and resistance. The mushrooming of Occupy protests, the Arab Spring, the mobilization of resistance against the Government of the Ukraine or in Hong Kong was heavily dependent on the resources provided by the social media. Many observers have concluded that in a networked world the social media possesses the potential to promote public participation, engagement and the process of democratizing public life. That the Internet and the social media are powerful instruments for mobilization of people is not in doubt. However, it is not its own technological imperative that allows the social media to play a prominent role in social protest. Rather the creative use of the social media is a response to aspirations and needs that pre-exist or at least exist independently of it. This technology ought to be perceived as a resource that can be utilized by social and political movements looking for a communication infrastructure to promote their cause.

Take the example of radicalized jihadist youth in the West. In many cases the Internet has been represented as a powerful technology that incites young Muslims to become radicalized. Often the term"sudden radicalization" is used to highlight the power of social media to swiftly convert otherwise confused young Muslims into hardened extremist jihadists. Yet there is considerable evidence to suggest that young Muslims who go online to visit jihadist websites have gone through a process of self-radicalization. They are already drawn towards radical Islam and are looking for a medium to express their ideals and interact with those who share their sentiments. What these websites do is to affirm, deepen or harden sentiments that their visitors already possess. Their experience of the Internet may encourage young Muslims to move in unexpected radical directions but these individuals have already developed attitudes that disposed them to embark on such a journey.

The relationship between the social media and radicalization is both an interactive and dynamic one. The social media provides a medium through which pre-existing sentiments can gain greater clarity, expressions and meaning. It provides a medium for the kind of interaction that can throw up new ideas, new symbols, new rituals and new identities. In this sense it has helped stimulate the emergent Western jihadist youth sub-culture and arguably its online expressions have exercised an important influence on its offline trajectory.

The Internet and Everyday Culture

The culture of everyday life has become entwined with the Internet. The flourishing of online dating offers a striking example of how the construction of significant relationships can draw on the resources provided by the social media. In many Western societies online dating has served as a provisional solution to the problems thrown up by a more individuated and segmented social setting. The stimulus for the cultivation of these online relations is the search for solutions to some of the problems confronting life in the offline world. However, the growing popularity of virtual encounters has had a significant impact on the way that men and women conduct their everyday affairs. The intermeshing of the virtual with the "real" is part of the reality of contemporary culture.

The influence of the Internet has been most significant in the way it has transformed the lives of young people. Their digital bedroom symbolizes a childhood that is significantly mediated through the social media, mobile phones and the Internet. Friendship interaction and peer-to-peer relations are increasingly conducted online or through text messaging. As with the case of political mobilization, the digitalization of childhood can be interpreted as a response to a pre-existing need for new technologies of interaction. The digital bedroom emerged as the outcome of the growing tendency to relocate children's activities from the outdoor to the indoor. Risk-averse attitudes which verge on paranoia emerged as one the defining features of contemporary child-rearing culture. Apprehensions about children's health and safety, particularly regarding sex predators have led to new limits imposed on children's freedom to explore the outdoors. This confinement of children indoors has been associated with the growth of a phenomenon frequently described as the bedroom culture. So the main driver of this process was not digital technology and the social media, but the prior development of an indoor childhood culture.

Bedroom culture and product of two interrelated

Bedroom culture is the product of two interrelated and sometimes contradictory developments. On the one hand the confinement of children indoors is the outcome of adult initiative. Media usage has become increasingly privatized and children play an influential role in the construction of the new media home environment. Many children's bedrooms are media-rich environments—a growing proportion of children have computers in bedroom with online access. Highly motivated to create a separate autonomous space where children can experiment and develop their personality, youngsters seek to evade parental control. The flourishing of bedroom culture encourages the privatization of media usage as young people attempt to forge a world that is distinct from that of their parents. Through pursuing the project of self-socialization, young people attempt to personalize their media to ensure that it directly relates to their interests. This project tends to be pursued in isolation from other family members.

The Influence of Popular Culture and Entertainment Media on Adult Education

The idea that popular culture and entertainment media influence us in both conscious and unconscious ways is not new.

New Media and Digital Culture

The study of new media is continually refreshed with objects, spaces, platforms and apps each seeking new users and niches: Snapchat, Medium, Tumblr, Instagram and Reddit operate alongside formerly 'new' media giants such as Google, Amazon, Facebook and Apple. Google and other engines and social media platforms are critiqued because their economic models of cheap aggregation and 'free' services in exchange for personal data have disrupted existing media industries such as music, publishing and news but also the taxi business through Uber and ola Personal data not only feeds Spotify and other systems that recommend taste and cultural preference, but also Nike fitness regimes and Apple health checks.

The program consists of three closely connected areas:

• Critical new media theory focuses on digital culture theories and histories and on the disruptions and transformations in culture and society brought about by social media, digital objects and devices, web and mobile cultures, locative and geo-media, ubiquitous computing, and digital aesthetics. Some of the topics are new artistic and cultural forms such as the digital book, smart cities, media , etc.

• Digital methods explore the possibilities for research into online-data cultures and in particular the potential of platforms and search engines for cultural, artistic, and empirical research interventions. Digital methods focus on 'digitally born' objects like hyperlinks, tags, 'like' buttons and tweets, as well as on specifically digital methods, techniques and strategies such as folksonomies and crowdsourcing.

• The program also comprises a strong Digital skills component, because knowledgeable experience with web culture and the ability to study and use web based applications like blogs, wikis, and software tools are considered critical skills for academic digital media experts and researchers.

Cultural Tools for Total Transformation of Our World

We have been commenting on the Internet and its specifics for a while. But for now I just want to comment on the wonder of it all. That in our hands we have a tool that could allow a total transformation of our world, by first transforming our values through visceral experiences and real-time sharing of information. The Planned Parenthood debacle (and the response to it) is an incredible example of this in action.

DIGITAL CULTURE AND SOCIAL MEDIA

A Brief Overview

If you are anticipating a roadmap of neat, organized plans for how the evolution of culture on digital platforms will unfurl. Lively discussion of how we define digital culture and what we might expect from it as it emerges in online spaces, mobile apps and platforms.

Additionally, this chapter includes a breakdown of the roles social media platforms may play in influencing culture.

Let's begins with a definition of "digital culture" that comes from the media studies portion of mass communication literature. **Media studies** refers to the broad category of academic inquiry analyzing and critiquing the mass media, its products, possible effects of messages and campaigns, and even media history. We can then continues with a deeper discussion of identity in the digital age and covers privacy and surveillance as well as the praxis of digital culture as defined by scholars. The term "praxis" here refers to how a theory plays out in actual practice.

Let'ts now identify different levels of culture (a concept borrowed from anthropology) as they relate to cultural products reaching audiences through digital mass communication channels. In other words, we ultimately answer this question: If we take existing theory for describing the levels of culture and apply it to digital culture, what are some immediately recognizable traits?

Digital Culture

Scholars argue whether we can understand what the spread of digital networks will mean for relatively well-established cultures in the tangible world, or predict with any certainty how cultures will evolve on digital platforms. There are two basic schools of thought. The first argues that existing cultures might find themselves essentially recreated in digital form as more and more life experiences, from the exciting to the mundane, play out in digital spaces.

It is worth noting that there are also niche fandoms that probably would not exist without the aid of digital networks. With virtually unlimited communication space, there is room for incredibly and they are not always socially positive communities. In many cases of hyper-specific fandoms, it is difficult to argue that these cultures existed in the physical world and simply "moved online."

Being digitally networked is what makes it possible to find people with particularly narrow shared interests, for better and for worse.

Digital Dynamic

Even with the presence of niche online groups, digital culture cannot currently be separated from the influence of physical-world cultures. We can say two things about the relationship between online and physical-world cultures at this time. First, the growth of interaction on digital networks influences "traditional" cultures. Second, longstanding cultural traditions are influencing digital culture as it takes shape. The ethics and norms established in the physical world shape our views about behavior and values in digital networks. The term **norm** refers to a behavioral standard. Mutual influences of what is considered "normal" in online behavior and well established physical world norms are emerging in a dynamic fashion. Sometimes they clash.

Online Dating

When online dating first became available, it was often compared to posting and perusing digital personal ads. This was a cultural perception based on previous experiences, behavior and expectations from a pre-Internet culture. Whatever it may be in a given culture, sexual morality still exists, even if new technologies make hooking up easier and new capabilities challenge old norms of what dating should be. This is the dynamic at the heart of this chapter. Digital technology can influence knowledge, beliefs and especially practices around dating.

Digital Individualism vs. Privacy

Should Google protect your searches and refuse to divulge information about your habits to governments, even if they share that data with other companies for marketing purposes? Should Google give you a way to hide your online activity? Is there a way for the liberty-loving Southeast Asian to have his privacy protected while still enabling Western governments to watch out for terrorists? These questions relate to larger issues of freedom and individualism in digital culture.

Throughout its history, the United States of America has taken pride in its First Amendment and the rest of the Bill of Rights as

guarantees of liberty. After the terrorist attacks of Sept. 11, many Americans accepted new levels of scrutiny, particularly in digital environments. Support for strong leaders increased. Concerns about the global rise of authoritarianism have people questioning government surveillance and corporate surveillance as they may limit our ability to engage as individuals in digital culture.

Online Mass Communication

Our experience with the anarchy of online mass communication platforms is quite limited. As we learn what government surveillance and corporate invasions of privacy are capable of, it may continue to deeply affect our physical world behavior. We have incredible freedoms and amazing digital communication capabilities as individuals living our lives in the new digital culture.

Internet and Smartphone

We can use the internet and smartphones to help us to get questions answered and to draw attention to ourselves in good ways. We can coordinate with others for fundraisers and to have parties. Digital communication networks are amazingly sophisticated tools that can help us connect as individuals to form groups to celebrate all sorts of interests, political and otherwise.

Capitalism depends on risk-taking, and if you kill risk-taking online, you have hindered the entrepreneurialism that the network society offers. We scholars will study for decades to come how individual behavior changes and how relationships morph in a digital culture that discourages behavior we want to keep private while simultaneously encouraging levels of sharing that border on exhibitionism. How can we maintain privacy and gain attention, which is so often the currency of the open Internet? This is an interesting dilemma that arises in an individualistic digital culture.

Post-nationalism

Post-nationalism is another aspect of digital culture that notes in his article. It may seem unrelated to our previous discussion of individualism and privacy in digital culture, but in fact, it is an analysis of the ways individuals represent themselves online.Most simply, **"post-nationalism"** in digital culture means that one's country appears to matter less as an influence on behavior and values online than it does in the tangible world, perhaps because we can be free of our national identities when engaging in digital networks with people from around the globe.This does not mean that we should expect to see an end to nationalism in the tangible world. Quite the opposite seems to be true: As post-nationalism appears in digital spaces. It might seem odd that people drop their nationalism online but demand it in physical spaces, but if you look at the way culture is expressed online, it is clear that for many people their nationality has little to do with their online identities.

For example, your country may be important to you, but it may not be one of the ways you define yourself in social media environments. You can love America without talking about it all of the time on Facebook or Twitter. Remember as well that national boundaries may be felt more readily in the daily lives of Africans, Asians, Europeans and others living in nations that are geographically smaller, more tightly packed and culturally distinct. In digital spaces, these cultural differences can evaporate.

Although war and immigration are highly influential on the current cultural climate in the physical world, the perception of evaporating culture in networked spaces may help drive the sense that physical world cultures are being threatened.Recent political developments, however, make it somewhat more difficult to think of digital culture as post-nationalistic given the rise of online nationalism — particularly white nationalism in Europe and the United States. White nationalism is a brand of nationalism related to white supremacy, but it is an identity connected to the nation-state nonetheless. A nationalist's primary modus operandi in digital culture may not reflect what nation states ultimately become in the 21st century, but rather what they wish it were. Even so, there is evidence that some factions will use digital spaces to promote a return to nationalism.

Does this mean that post-nationalism in digital culture is a false notion conceived in the early 2000s that has no bearing on culture today? Perhaps, but it is more likely that we are seeing a backlash against the rise of a global post-nationalist space online.

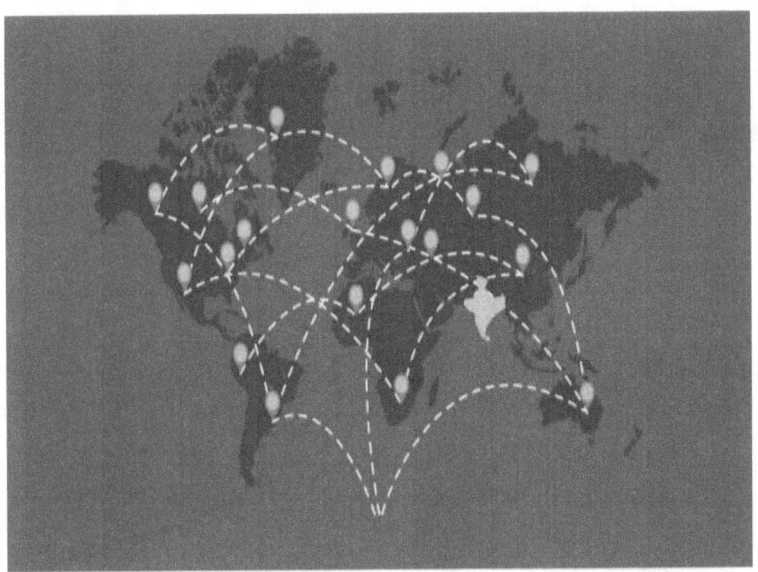

Globalization

Behaviors, interests, and relationships cross international boundaries. The economic structure of digital networks, including the mass media system, is global. For example, multinational conglomerate corporations tend to dominate the media industry show that most mass media companies fall under the ownership of large corporate firms. It is not accurate to say this represents all media or that "the media" are being controlled, but it is accurate to say a significant level of influence can be attributed to a handful of media corporations in most developed parts of the world.Mass media consumers should be aware of the environment in which media products are produced, but this is not to say that the globalization of mass media is always a negative thing. When it comes to culture, globalization has its supporters. The music comes from Korea, but the fan base is spread worldwide, and the site can reach a global audience only because of the global nature of digital networks. It works only because computer servers are connected by wires all over the globe to make this bit of culture, like many others, available to the entire globe. had the formula, we'd include it here.) All we can say at this time is that you can reach the world with any online message and, for whatever reason, some things are globally likable and "shareable."

A Place Called Gangnam

Humanity's recently developed ability to develop a globalized point of view and to establish a common digital culture is the reason you have heard (and likely tired) of "Gangnam Style."

Ironically, PSY, who performs the song, is kind of an anti-pop star within Korea. The song makes fun of the country's higher class, a conspicuously wealthy subculture from a place called the Gangnam District

Digital Culture in Practice

Participation means that every individual will have the ability to contribute to online media. Professionals and amateurs will work together much more often than they did on "legacy media" products and projects.Because people do not want to work for free, they will not flock to an online platform simply because it has been opened up for contributions. If anyone could build a Facebook, there would be hundreds or even thousands of competing platforms. Participation is an essential part of digital culture. It can be easy and fun to do it for free. If you want to make a career out of it, it takes professional-level commitment, and the resulting content often favors what is popular and emotionally gripping rather than what is

A good example of remediation is taking classic movies or video games and showing them to young people to record their reactions for YouTube. Do it yourself by combining elements found elsewhere. Much of digital culture is an amalgamation of existing content and new cultural work being done at home by people with amateur skills and affordable but capable tools, such as smartphones and tablet computers. Even basic tools are quite powerful. Smartphones come with front- and back-facing cameras as well as HD-quality video. The computing power of a smartphone is more powerful than a mainframe computer was 70 years ago. Independent producers have video and audio editing software options and can create professional looking, popular media products on their own with little formal training.

Professionalism

What is formal training for, then? It prepares you to transition from making professional looking and sounding media products once in a while to consistently making professional quality media. Formal training prepares you to think strategically about where industries are going so that you know not only how to make mass media products but where to place them and how to use and possibly develop your own communication platforms. Planning for multiple media shifts and seeing digital cultural trends as or before they emerge requires an education in more than the tools and tricks of the trade.

Levels of Culture in Digital Media

It is more accurate to say that the knowledge, beliefs and practices of a massive group of people at a certain time and place. Three levels of culture exist in anthropology literature, and they apply to the ways culture is expressed in the mass media. The three levels of culture are personal culture, group culture and common culture (similar to pop culture).Any kind of culture, whether it is personal, group or common culture, relies on shared knowledge. There must be shared experiences and shared stories about those experiences for us to have a common culture. If we did not have shared experiences, cultural references would not make sense. Thus common culture can be arrived at when individuals and groups tell the same stories, or when mass media reach mass audiences with the same messages at the same (or about the same) time.

The more people who know about a song, film, work of art or event with cultural significance, and the more information that they know about it, the more likely it is that event will become part of the common culture. The mass media influence common culture, although it is not correct to say that they directly shape it. There are many other institutional influences on common culture such as governments, churches, families and educational systems.

In fact, messages in the mass media may not be as influential now as they were in the mid-20th century when millions of people watched the same TV shows each week at the same time and read the same major metropolitan daily newspapers and national magazines. Demassification has affected the ways common culture is established and fed.

The mass media influence may have less power to influence common culture directly, but it is still relevant. Think about any major global news event of the past few months. When an event is big enough that it is shared across all media platforms, especially cable television, broadcast television and social media channels, it can form a piece of common culture.**Group culture** is what we used to refer to as a "subculture." It is the knowledge, beliefs and practices of a subset of people considered to be part of a larger culture. Group culture is distinct in some ways from the shared, broader common culture. Group culture might center on religious beliefs and practices, ethnic norms and interests, or food, music and other forms of material production.

You have a say in defining your **personal culture**— the knowledge, beliefs and practices held most dear to the individual. You may find yourself identifying with many group cultures or taking most of your interests from the dominant common culture. Do you take your cultural cues about what to think about and talk about from television, social media or small group cultures with which you identify? This much is your prerogative. You can choose your personal culture. It is based both on what you believe in and what cultural products you consume.

There is a common culture in America, but there is no single, dominant, common culture across global digital networks. There may be a tendency for people to believe that the group cultures they interact with most often online constitute the "real" digital culture, but as yet there is no clear consensus about what our shared digital culture is or even if we will develop one.

Algorithms in search engines and social media platforms determine much of what we find when we search the internet and what we see when we look at news and information feeds from our friends. Do algorithms constitute common culture? They may shape it, and they may be influenced by user preferences, but they are not always designed for truth, accuracy or information literacy. They are most often designed to give consumers whatever makes them consume more of what the platform wants them to consume. Google usually wants you to spend money with its advertisers. Facebook wants your time and your data so it can sell your information to third party advertisers.

What shapes digital culture is often in a "black box": It is the proprietary information of very large corporations, and the public may or may not have access to the code. Even if we did have it, it would be difficult to explain exactly how algorithms work. There are times when the corporations that deploy algorithms seem surprised by how they function in the hands of massive numbers of users.

Major events that cut across algorithms and show up on almost everyone's news feed and in almost everyone's search results are still likely to have an impact on common culture. Major events are likely to shape personal, group, and common culture if they are significant enough. What kind of cultural impact does a given event have? It depends.The impact of a school shooting near Miami might be felt differently in Florida than in California because of proximity and because the gun laws in each state are quite different. In other words, something can enter the common culture but still be perceived quite differently by individual members of the public.

Norms

By now you should understand that the cultural impact of messages in the mass media at each level — personal, group and common culture — is related to the shared knowledge that existed before the event.Events will usually be interpreted differently by individuals within a small group culture, depending on an individual's beliefs about and personal experiences with the issue at hand.A person's response to current events as they appear in the mass media is also related to the existence and strength of shared beliefs about the way they think things ought to be. We call those beliefs **cultural norms**.There is no single, agreed-upon set of norms that everyone in a given group culture adheres to. If you have lived your whole life as part of the dominant culture, and you do not recognize the existence and struggle of various cultural groups, it can be difficult to recognize reactions in digital media spaces that do not relate much to what you see in your physical world. Conversely, if you have grown up being oppressed as part of a small group, you may find it hard to understand how others identifying with the dominant portion of a common culture can miss the cruelty present in some cultural norms they don't think twice about.Exposure to other groups' cultures in a network society can bring about both greater understanding and greater anxiety. This is something that will be worked out, for better or for worse, over the next several decades as digital culture evolves. Figuring out how groups with different cultural interests, norms, and values can get along while being constantly exposed to one another's views in the free-for-all of network society is the challenge of emergent digital culture.One response is to run to echo chambers, to partisan spaces that feel safe for certain group cultures and for our personal cultural beliefs and priorities, but this practice can only deepen the divide between cultural groups.

Common culture and network society

In the early years of working to establish a common culture in the network society, we have managed to inundate ourselves with information from all manner of cultural groups and to isolate ourselves from views that contradict our own group cultural norms. This is anarchy. This is culture without a strong social structure to hold it together.

The question facing mass communication scholars that members of our common culture also face is whether the institutions of the physical world can or should try to control how digital culture is shaped. You have the power to decide if digital culture should be regulated and how. This may be the most important civic responsibility you have, but it is also a matter of cultural power.

Social Media and Social Capital

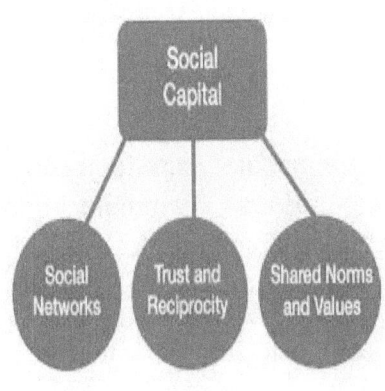

That means every individual with a computer or a smartphone has the potential to disseminate messages that influence broader society. Think of #Ferguson protests in the summer and fall of 2014. Think of the way candidate Donald Trump by passed mass media outlets to reach voters and to set a separate news agenda in 2015 and 2016. Individuals and small groups are now able to coordinate and to lead social movements using networked communication technologies.

You have probably heard the term "social movement." In a sense, a social movement is a change in society brought on by communication. What is different about the world of networked communication is how interpersonal messages and message campaigns can shift in an instant to being mass messages or massive campaigns. This makes digital networks battlegrounds because networked public communication platforms are centers of power now more than ever. This makes our communication system as ripe for abuse by outside forces as it is for use by legitimate citizens. Governments, corporations and rogue dictators all have an interest in learning our secrets, and they could potentially hold them against us.We cannot underestimate how important this is will be in the mass communication field. Individual, group and broader social secrets — including consumer behavior, political behavior and even personal thoughts and interests — are easier to discern and possibly manipulate than ever before because of the vast amounts of data collected about us from our social media and other internet habits. This can have a profound effect on our behavior and on our society, and we are not prepared as a society to defend ourselves against attacks.

Interpersonal communication, organizational communication and mass communication are separate areas of academic interest, as stated in the first chapter, but our ability as consumers and as producers to alternate from one to the other is as powerful as it has ever been. Being connected to each other almost at all times by digital networks creates the capacity for relatively quick mass social action. People are beginning to use this power to pull society in different directions. Large numbers of people can be organized and we could see social shifts and rifts develop more quickly than they can be put back together.

Participatory Media

A major shaper of culture and society is the news media. There will be separate sections on the evolution of news in later chapters, but in the context of digital culture, it bears noting that the role of news media within broader media landscapes is also shifting.Apart from the ability of social movements and cultural movements to arise and take shape on social media platforms, there is also the potential for public opinion to be influenced quickly and deeply when mass media outlets operated in the same digital networks as influential individuals and groups.

5. Social Media Trend And Internet

Social Media

Not too long ago management consultants told business owners that they needed a website to succeed in business. Now many recommend a website and a social media policy that includes Facebook, Google Plus, Twitter and YouTube.A business needs to be on all of these social media websites. They all do different things. One is fine, two is good, three would be adequate and being on all four would be just right.

Consumers, customers or clients investigate before they make buying choices. Every one of the four most influential social media websites has the capacity to influence buying decisions. A business has to be in the mobile online customer race. Not turning up to the event means that a business misses out on more and more cash.

1. It is a bad look not to be on a Facebook business page. Apart from the obvious inability to interact with a potential client, buyers think thoughts like.

2.Start with the basic Facebook start page and then always add extra features like header images with contact details clearly visible embed in the picture.(Be sure to include the company logo)Never just leave the basic sign up start page. Take an interest and show potential buyers that you are not ordinary.

3. Post something on the Facebook Business Page every day. Build a dialogue. Create a conversation. Get real life friends to become online Facebook friends in the form of likes. Get them to comment on the posts. Build a community of people who interact with the business on the business page. It is vital. An active up to date Facebook page shows that the business is vibrant. The comments can be testimonials. Buyers like favourable testimonials.

3. Google will visit a business Facebook page and gather all sorts of information that will affect the amount of traffic that is delivered to the page in response to online searches. Make sure that the posts have keywords that are appropriate for the business. "Coffee, Hair, Office Supplies ...."

4. Accumulate "Likes" but do not buy likes. People who sell "Likes" do not come online to post on the Facebook page. Search engines can see the fakes and will either ignore them or worse, penalize the pagerank of the business pages.

6. Link to articles, images and website pages that relate to the business. Remember that Facebook does not restrict the size of posts in the same way that Twitter does. Go for it. Use it. Posts beget comments and social media buzz

Using Twitter to grow a business.

1. Twitter is not a passing fad. It is a part of cyber life. It is a source of news and information. Often it is the source of breaking news that beats the major news businesses. If a business has "news" then Twitter is one place to use to ensure that the news reaches a lot of potential buyers - customers - cash.

2. Followers are great. Twitter people who look forward to a business post on Twitter are called "Followers". They want to hear from the business or its people. Just like real life, you have to talk to people you care about. If you don't, they will opt out of a Twitter feed and will they will no longer get marketing news. Keep the flow of information going. They signed up for it. Deliver it.

3. Follow your followers. Don't just send out Twitter messages. Receive them and respond. Sign up to receive Tweets from the people who were kind enough to "follow". They will have something to say to a business. They will give valuable clues to their likes and dislikes. That information is valuable and it is marketing gems that a business can exploit.

4. Use Twitter to post links to new, valuable and informative new content on the business news website.

Using Google Plus to grow a business

1. Google spent a lot of money to create Google Plus. It is brilliant. There is so much to do on Google Plus. There are "Likes" in the form of Plus votes, video chat facilities, advertising opportunities and many ways to interact and categorize contacts. Take the time to learn about Google Plus. It requires a Google account that also permits access to gmail which is also brilliant, indispensable and a valuable communications tool in its own right.

2. Post news about the business to everyone who has signed up.

3. Seek out and read posts by people who interact with a business on Google Plus. Their comments are valuable. They will appreciate the feedback and many will return the favor by becoming buyers, customers or clients.

4. Use Google Plus to have video conferences. Present news, new products and a weekly bulletin with news that customers want. It can be the communications hub of a business.

Using YouTube to grow a business.

1. You tube is the perfect place to host business news videos. Compile catalogs. Create videos (cheaply on an iphone) and upload them to YouTube on a channel linked to a business gmail and Google Plus account. Be sure to embed videos from YouTube to the business' website to reach more customers.

2. More people visit YouTube than any business website, so be where the people (customers) are (Constantly) .

3. Actually create content for YouTube. Do not just send junk up there.

4. A business will have a free talking sales person if they put well planned and produced advertising on YouTube.

5. It is a perfect medium to introduce a business and its people.

6. Create and post new video content weekly. It keeps people thinking about the business and its products.

6.Social Media Is The Present & Future Of New Media

Social Media Is the Present & Future of New Media

Even though I've since moved on to topics such as digital transformation, innovation, experience design, and corporate culture, As a digital anthropologist, I still track how people and their behaviors, values, norms, etc., are evolving as a result of social media. As you can imagine, this spans everything from politics to work to education to beauty and health and more.

How Social Media Has Evolved and Where It Is Headed

The early days of social media were a really exciting time. In Silicon Valley in the mid-2000s, Facebook opened to the public in 2006, Twitter appeared in 2007, and early social networks like Friend Feed and Friendster were still around. Some people saw the promise of social media but most of the world had

Plus, it was the beginning of entrepreneurship, at least in this era. Everybody in every industry was suddenly a social media pro. You had marketers, advertisers, coaches, you name it. Everybody latched onto social media because it seemed like it was the next gold rush.

The Rise of Live Video, Messaging, and More

Live Video. Live video is the next iteration of how people share themselves, as well as how they watch or follow others. In the beginning, social media was largely text and then evolved into imagery, and later, more sophisticated imagery. Video started with YouTube, Vimeo, and other early networks. Now, live video has turned people into real-time broadcasters.

Many of these people simply share who they are and what they're doing right now. Or they use live video to share happy moments. However, others use live video as full-on broadcast mechanisms to engage new audiences in ways that weren't possible before. These users are almost becoming like their own TV networks.Facebook has really demonstrated what's possible because it's such a huge platform where people have a built-in audience. The algorithm will definitely affect who you do and don't reach. At the moment, Facebook is putting substance behind live, so now is probably the best time ever to get into live video.

Messaging.

Messaging is the result of people's increasingly mobile lifestyle.Although Path came and went, you can still find messaging that harnesses the private, public, and other types of moments. There's this famous saying, "You live a private life, you live a public life, you live a secret life, and there are networks for all of them."

Augmented Reality.

Augmented reality goes back to the early days of mobile. It allows you to add a layer that really brings the real world to life. You could hold the phone over UPC codes in supermarkets, and product information appeared. QR codes also came out around that time, and others were experimenting with consumer applications, too.

Social Television.

Facebook Watch was just announced, while major companies including Snapchat, Twitter, and Apple are making huge bets bringing television to mobile and some of the social platforms.Social TV is the next big thing, Brian says. It was born out of what people did anyway: watching television and going on social media to participate in the conversation. Content playing directly on the network will further integrate the experience.In early attempts, dedicated social networks had content as the mainstay, and people engaged around the content. However, better opportunities come when you engage with your social or interest graph and introduce content into that mix. Everybody's starting to bid on and create content. It's a matter of time until we start to see what Netflix, Amazon, and Apple create.

Artificial Intelligence.

Anytime there's a new technology, marketing tends to put that technology in the portfolio of classical marketing. Brian has seen this with social media, the Internet of things, everything. He believes artificial intelligence isn't something that should simply be added to the mix. Marketing is ripe for innovation and disruption. So much is possible when you start thinking differently about data, platforms, and opportunities. Artificial intelligence will make certain things better and traditional marketing even worse. As long as people's minds are open to doing things in new ways, artificial intelligence will make this generation of marketers the most valuable breed. When individuals think differently about their approaches and build on that, that's when they win. **The world of social media is constantly evolving, from months, days to mere minutes! Over the last decade as it has come to be the dominant platform in the world of media, investments have gone up and businesses are seeing the true value of staying social with their customers.**

BLAST FROM THE PAST

2010 was a tipping point for social media. News and entertainment related content gained heavy traction with users, giving rise to the BuzzFeeds of the world.Facebook took over Google as the dominant platform for digital media ads and it has only strengthened its presence. Giving rise to dedicated social media campaigns that used to only take a backseat to traditional marketing efforts till.E-commerce came up in a big way around the world and people understood the benefits and convenience in ordering things right from the comfort of their homes! As internet penetration and infrastructure improved, businesses made social a priority. While the benefits were countless, suddenly everyone in the industry became a social media pro overnight, making it imperative to actually gauge the quality of content and marketing efforts, and the way content is served to consumers.

THE PRESENT IS ALL ABOUT

Today, success is all about grabbing consumer attention and reaching them through content that speaks to them. No more hard-selling, click-baiting and asking your fans for likes, shares and comments! The key is to think like a marketer, solve problems, generate results and lead customers to initiate a sale through storytelling content.

The focus is on:

- ORGANIC REACH

Facebook's recent changes amid privacy concerns has led them to make it harder than ever for branded content to stand out organically. Even when it comes to paid campaigns the way their algorithm works means that content needs to urge genuine discussions between fans with longer comments leading the way.

- STORIES

People are consuming content on the go, giving rise to bite-sized videos across platforms. Stories across Instagram and Snapchat platforms are seeing the highest growth through them. The future is certainly based on these shorter videos and contrary to popular belief there has been an increase in the consumption of vertical videos with more sound turned on.

- CONTENT CREATORS

Influence now takes center stage. While traditionally word of mouth was considered the holy grail of effective marketing without massive spends, social media influencers and content creators enable the same today. They have loyal followings big or small that listens to their suggestion, create meaningful dialogue, rapidly accelerate growth and are seen to me more effective than just seeding content.

WHAT'S IN STORE? THE FUTURE OF SOCIAL MEDIA

Video has been the reigning king of content and in the coming years it will further evolve in new ways yet unfounded! Live video is the latest of iteration of it and is here to stay. As organic reach goes down, marketers have seen the huge potential of live videos reaching fans organically, you'll be hard-pressed to find any other medium that delivers so much with so little investment.Social television is also coming up in a big way with the announcement of Facebook Watch and upcoming developments across the board by Snapchat, Twitter, Instagram and all the other heavyweights in the world of media and tech.

Politics and social unrest have been around since man first gathered into groups in order to survive. Someone always wants to be in charge, and someone is always unhappy with the way things are. Leaders have always used private or public communication tools and propaganda to gain or maintain power, and those who oppose them have always used the same to overcome them when necessary moved things a step further in early American history. Today, those same ideas have been continued with Twitter, Facebook, and blogs.The difference today, with instant communication possible on a global scale through email, IM, and social media, is that these things happen much faster, and often much more effectively. With the near ubiquitous nature and adoption of social media and online information dissemination in general, being able to monitor that activity is more important than ever for both sides of the equation – the leaders and the people.This can be a liberating force for good and freedom, but there is also a dark side as authoritarian governments – and frighteningly, many who claim to be for freedom – use the same tools to monitor their citizens for any dissent in order to clamp down and maintain control.

The Good

The new mass communication tools have seen their share of successes in the last two decades. The ability to communicate and organize quickly via text messaging led to the ouster of corrupt Philippine president Joseph Estrada in 2001. Text messaging also helped get rid of Spanish Prime Minister José María Aznar in 2004.

Text messaging combined with Facebook and Twitter contributed to the downfall of the Communist Party in Moldova in 2009. Perhaps the most well-known use of social media in a political sense came in the form of Barack Obama's election as President of the US in 2008. Coming at the perfect time regarding digital and social media's evolutionary timeline, the combination of Twitter, Facebook, websites, blogs, and online news sources played a huge part in the election of an unknown, one-term state senator who also happened to be the first person of color elected to that position in the nation's history.

Since then, social media and other internet outlets have played a regular, and often crucial, role in politics, including local, state, and national levels, and in the international political discourse as well.

The Bad

Before the new age of communication, not all movements, uprisings, or instances of political or social activism were successful, and many were brought to a quick and violent halt by those in power. Some things never change. There have been some recent historical cases where despite the ability of people to communicate and organize quickly for a cause, the results were devastating for them.

Authoritarian governments such as China have long held tight controls over internet access for their citizens. Unfortunately, even nations that claim to be "democratic" (a misnomer if ever there was one) such as the US and the United Kingdom are guilty of curtailing their citizens' freedoms.

Thankfully, most of us don't have to be quite as concerned about many of these issues. We are instead focused on more sensible and practical business and political uses for social media, blogs, email, and sms. These more extreme instances of digital communication social media monitoring, however, do much to emphasize the importance and usefulness of such measures in our daily personal, business, and political activities.

It is without question that both proper political activity and propaganda will be increasingly influenced and enhanced by our digital means. Elections are now, in essence but with some limitations, held online before voters ever reach the ballot box. Good or bad, this is now reality, and needs to be handled as such.

While private business has been gaining steam for years in analyzing the sentiment of their brands and customers using social media monitoring software, the practice is now increasingly used by political campaigns as well. What has been considered an indispensable marketing tool for companies for several years has become an indispensable tool for political campaigns and activism as well.

The Strategy

Fortunately, the ability to perform this kind of sentiment analysis and online monitoring isn't limited to national security agencies. Any local political campaign can obtain a reliable sense for the community's thoughts and feeling issues or candidates with a readily available and efficient piece of social media management software.It's important to not only be able to monitor Facebook and Twitter, but other social networks, blogs, websites, and online news sites as well. The political discourse occurs in every nook and cranny of the internet, and confining the mined information to one or two networks will simply not yield a complete picture of the wisdom, or lack thereof, of the crowd.

Wikipedia defines Social Media as "media for social interaction, using highly accessible and scalable communication techniques. Social media is the use of web-based and mobile technologies to turn communication into interactive dialogue." By this definition, the first online social media platform was Usenet, an early online bulletin board. Social interaction in this medium was a kind of structured anarchy, not for the faint of heart or thin of skin. Over the past twenty years or so, social media has grown and matured, as advances in technology have enabled increasingly sophisticated platforms.

Then, as now, social media served as a way to democratize the sharing of content, enabling individuals with limited online influence to reach a wide audience. Think of it as a democracy in which every participant is running for office, and voting on every other participant. Stories show how many up or down votes they have received, but not who voted either way. Submissions reach the front page based on a combination of overall positive votes, ratio of positive to negative votes, and the age of the submission. Comments are sorted, by default, by the popularity of the comment, and are threaded. By utilizing a system where users are recognized for their track record of contributions, this introduces an element of meritocracy to the democratic process, and encourages users to be mindful of their behavior. Reward systems also help build member loyalty, as it offers tangible and measurable feedback for their positive contributions to the community.

Social Bookmarking: Users Become Curators

This variation on the social media sharing sites turns members into curators of content; the focus is on allowing the user to collect and share favorites. It offers users the ability to organize their bookmarks in categories and tag them by keyword, encouraging them to use the service as a personal organizer of sorts

The Democratization of Content

In the early days of the social web, the popularity of content was determined mainly by a small number of key influencers. In this atmosphere of mostly one-way communication, even those popular sites that did offer comment functionality would do little more than moderate the submissions. With the advent of social sharing, the web shifted from a mostly broadcast medium to a platform for conversation, and with the addition of crowd sourced rating systems, virtually any story now has the potential to become popular.

The Rise of the Semantic Web

As these technologies evolve, companies that use them in non-threatening and beneficial ways will open up new avenues for success; transparency and demonstrable good intentions, combined with responsiveness to privacy concerns, are the keys to earning the goodwill of the users.

The Human Connection

No matter how much technology is thrown into the mix, the functional approach should always be focused on creating and nurturing human connection. Consumers develop their strongest brand allegiances based upon positive interactions, even if those initially come about due to negative experiences. Perhaps the biggest challenge we face is to avoid letting technology lead us into impersonal and automated interactions. And perhaps the ultimate goal of this all is to embrace technology that breaks down barriers and connects people with other people.

Social media is changing at a rapid pace.

For example, the content format Stories was introduced on several major social media platforms slightly more than a year ago. Now, it's one of the most popular content individuals are posting. More than 300 million people are sharing stories on Instagram and WhatsApp separately every day, and 70 million people were posting daily to Messenger Day just six months after its launch
Due to its popularity, many businesses are now posting stories regularly, too.

Top 10 social media trends to know for 2018

5. Chatbots are becoming the norm

6. Businesses can no longer afford to ignore social customer service

7. Organic reach and referral traffic are plummeting

8. Video is still the most popular content type

9. User-generated content can help drive reach and engagement

10. Businesses are pouring more money into social ads

There are now almost 2.5 billion social media users

Social media is one of the best ways to reach your target audience because of its sheer size and the amount of data available.

Here are the current user base of the six major social media platforms which are as follows

- Facebook: 2.07 billion monthly active users
- Instagram: 800 million monthly active users
- Twitter: 330 million monthly active users
- LinkedIn: 500 million members
- Pinterest: 200 million monthly active users
- Snapchat: 178 million daily active users

You can see the growth of the respective social media platforms in the following few slides:

Social media is going mobile

We are becoming glued to our smartphones.

Facebook studied the behavior of 100 people while they were watching TV at home. 94 of them had their smartphone in their hands while watching TV. And one of the top reasons they look away from the TV is to use their smartphone.

More and more people are also using social media on their smartphone. In fact, the majority of social media users are using the apps via their smartphone:

- Facebook: 94 percent of its monthly active users

- Instagram: Because it's a mobile-first app, I think it's safe to assume most of its users use the app on mobile

- Twitter: 82 percent of its monthly active users
- LinkedIn: 60 percent of its unique visitors access LinkedIn via a mobile device
- Pinterest: 80 percent of its traffic comes from mobile devices

- Snapchat: Snapchat only has a mobile app and no web application yet.

On top of these statistics, found that 78 percent of social media time is spent on mobile devices.

This trend increases the importance of creating mobile-friendly or mobile-first content.

.

Social messaging will greatly benefit businesses

People used to communicate with businesses via the phone, then emails, and then social media. Now, it's social messaging.Facebook studied the messaging behavior of 12,500 people across 14 markets to understand consumers' growing preference for messaging businesses. They found that messaging is helping businesses connect with their customers more than ever. Consumers use messaging to ask businesses questions, make appointments and purchases, and provide feedback.

Here are some other findings:

- Fifty-six percent would rather message than call a business for customer service

- Sixty-one percent likes receiving personalized messages from businesses

- More than fifty percent prefers shopping with a business they can message

Consumers' preferred way of communicating with businesses has always been shifting. Now, they are shifting towards social messaging. Is your business ready to receive your customers' messages?

Businesses can no longer afford to ignore social customer serviceCustomer service used to be private conversations between a customer and a customer service representative. Social media has changed that entirely. Businesses are pouring more money into social ads.Social media ad spending has also been on the rise, growing more than 20 percent annually.

What other social media trends have you spotted?

As the major social media platforms continue to grow and introduce new features, the social media landscape will only get more exciting. I hope by sharing these ten current social media trends, you can stay ahead of the curve and succeed on social media.

There are now almost 2.5 billion social media users

Social media is going mobile

Social messaging overtook social media

Social messaging will greatly benefit businesses

Chat bots are becoming the norm

Businesses can no longer afford to ignore social customer service

Organic reach and referral traffic are plummeting

Video is still the most popular content type

User-generated content can help drive reach and engagement

Businesses are pouring more money into social ads

In the world with over 70% of internet users' active on social networks, who spend at least one hour a day on average on those social networks, we have to conclude that social networks have become a sort of reality in which people communicate, interact, and obviously trust. We also have to be aware that over 60% of those users access social networks via mobile devices, with strong indicators that this percent will only increase in the future years.

Thanking You Notes

To all those geeks and lovers of social media who got this book and read this book to conclude it with the facet of ever changing community of social media.

This is just the start with new and new media emerging every day that are immense ways of getting influential in our lifestyle.

So Hope everyone will find a new trend in their social media profile or can influence some trends till the next edition of the emerging trends hope you guys will keep influencing other trends and also other popular culture that will help to convey a meaning also to create a popular culture or trend in social media.

Adios! Folks keep reading the content and can always keep in touch via email for any enquiries or discussions can connect via email @ nimesh22263@gmail.com

--------------------The End-----------------

www.ingramcontent.com/pod-product-compliance
Lightning Source LLC
Chambersburg PA
CBHW020542220526
45463CB00006B/2163